With Reason and Vision
Structured Dialogic Design

Based on the work of Dr. Alexander N. Christakis
& the Members of
The "Institute of 21st Century Agoras"
Compiled & Edited by Dr. Kenneth Bausch

Ongoing Emergence Press

With Reason and Vision
Structured Dialogic Design

Bausch, Kenneth C.
With Reason and Vision: Structured Dialogic Design
pp.
ISBN-1517757649
1. Systems Science   2. Social Systems Design 3. Conscious Evolution

Cover Photography: Maria Kakoulakis – Independent Journalist
Cover Design: George Kakoulakis – Graphic Designer
Illustrator-Graphic Designer -- Alexandra Mousourouli

Ongoing Emergence Press
Cincinnati, OH 45274

Structured Dialogic Design
A Science with Wings

It enables people
>To voice their aspirations
>To pool their wisdom
>And unleash their power
>To create the future.

It provides the backbone on which
>The sinews of the people
>Can reinvent democracy.

Vision is evident in all three stages of Structured Dialogic Design, especially as participants voice their aspirations, organize their vision, and identify the vision's most influential components. It is visible in their identifying obstacles to that vision and finding the most influential barriers. It is also evident in their identifying ways to overcome those obstacles and devising an effective strategy for overcoming them.

Reason is also present in all three stages:
>As intuition and abductive logic in every step of SDD
>As inductive logic in the testing of axioms, laws, and practice
>As systemic modeling in its structure and methodology
>And as deductive logic in the coherence of its theory and practice.

This careful articulation of vision and reason in the pursuit of effective dialogue results in elegant colaboratories, jubilant conclusions, and committed action.

Acknowledgements

Many people contributed to the construction of SDD as related in this Booklet. First among them are Aleco Christakis and the four giants who stand beside him: Constantine Doxiadis, Hasan Ozbekhan, Aurelio Peccei, and John Warfield. In the recent past, members of 21st Century Agoras and the World Futures Center participated in the extended online discussions which supply the essence of this booklet. Among them, are Kevin Dye, Tom Flanagan, Yiannis Laouris, Peter Jones, Gayle Underwood, Reynaldo Trevino, Norma Romm, Janet McIntyre, Jeff Diedrich, Diane Conaway, Jacqueline Wasilewski, LaDonna Harris, Laura Harris, Paul Hayes, Marios Michaelides, Maria Kakoulakis, Ioanna Tsivacou, Roy Smith, Heiner Benking, Larry Fergeson, and Craig Lindell. Closer to home, special thanks to Maria and George Kakoulakis who supplied great poetry and artwork for the cover. Very personal thanks to Gayle Underwood who enabled the Obamavision section and solved graphic and formatting difficulties.

To all these people and those of earlier times who contributed their wisdom in laying the foundations for Structured Democratic Democracy, we owe profound thanks.

Table of Contents

With Reason and Vision

When a group of people get together to find a way to deal with a situation, in a neighborhood say, you will find people with different backgrounds, different cultures, different prejudices, and different perspectives. Some will be visionary and thoughtful; others will consider them to be dreamers. Some will come in sure of what should be done and try to get the group to do it their way. Some will want to talk things out and get everybody's opinion; others will consider that a waste of time. Often such meetings end up at a dead end; sometimes a bunch of ideas are put up and participants vote on which ones they think is most important, and decide to follow the most popular idea.

Even in day long sessions with competent facilitators very little real work is done. The resultant plans are made without sufficient group consideration and put together by "experts." Such half-hearted efforts to involve a community seldom if ever get off the paper they are printed on. As a rule, such efforts are supplanted by heavy top-down planning

One young man, Aleco Christakis, observed this while working in city planning efforts in the1960s and 70s. He saw the sorry effects of excluding stakeholders from real decision-making, but also saw the seemingly insurmountable odds against full citizen involvement. He has spent his life devising ways to make bottom-up planning possible and feasible. His efforts along with those of others have now resulted in a science of dialogic design.

Evolution of the Science

Sciences develop over time and are the products of many collaborating individuals. Sometimes there is a unifying figure (or figures) who pulls together the lessons of several collaborators. In the development of the science of dialogic design (SDD) the unifying character is Alexander (Aleco) Christakis. The main characters, other than Aleco are Constantine Doxiadis, Aurelio Peccei, Hasan Ozbekhan, Erich Jantsch, W. West Churchman, John Warfield, and Kevin Dye.

Aleco returned to Greece in 1965 with his newly minted PhD in nuclear physics from Yale University. He was enlisted by Constantine Doxiadis and charged to supply scientific rigor to the science of human habitations (Ekistics). Doxiadis was a foremost architect and urban designer of his time. With Ekistics, he was striving to provide a scientific foundation for his work in designing and producing desirable living spaces for human communities. In addition to his urban planning projects, Doxiadis also hosted global think tank discussions with major luminaries of the time including Arnold Toynbee, Buckminster Fuller, Margaret Mead, and many others. During two-week cruises among the Aegean islands on Doxiadis' yacht, they would discuss the major problems of the postwar world and propose solutions to them. Docking at the isle of Delos at the end of the cruise they would compose the conclusions of their deliberations.

Aleco was striving to apply the rigors of mainstream science to the urban design problems he was working on. He was also participating in and observing the discussions aboard the yacht. From his time with Doxiadis, he came away with an abiding drive to develop a conscientious community planning methodology. He also came away with some critical conclusions and observations:

- Human community planning has to deal with all aspect of human life.
- Planning requires tools and methods beyond those supplied by physical science.

- Adequate communication when dealing with global problems requires a shared language not supplied by various scientific, historical, and philosophical disciplines.

While working with Doxiadis, Aleco developed a close working relationship with Hasan Ozbekhan, the author of an important article entitled Toward a General Theory of Planning (1968).

Aurelio Peccei, an Italian industrialist, was concerned about the imbalance of wealth, technology, and opportunity in the postwar era. In his book, *The Chasm Ahead* (1969), he detailed a number of problems facing the postwar world. In the late 1960s, he was traveling the world trying to persuade leaders that we were facing an unprecedented global crisis. By 1969, he was seeking an effective methodology to tackle the issues of the problematique, which he called "a tidal wave of global problems."

To that end, he sought out Hasan Ozbekhan who was considered one of the most prominent planning theoreticians of the time. Peccei and Ozbekhan became very good friends and during one of their meetings, Ozbekhan proposed to Peccei that he adopt the systems approach to deal with the problematique. The executive committee of the Club of Rome commissioned Ozbekhan to write the prospectus for the Club of Rome in 1969.

Hasan working together with Aleco, Erich Jantsch, and Aurelio Peccei produced the prospectus in 1970. It was titled *The Predicament of Mankind* and called itself a quest for Structural Reponses to Growing World-wide Complexities and Uncertainties. It broke those challenges down into 49 continuous critical problems in areas such as population growth, poverty, warfare, education, prejudices, environment, and value bases. Some of these problems can be measured to some extent with quantitative tools, while some are charged with social values and cannot be objectively quantified. It stressed that none of these problems can be addressed solely on its own terms and a balanced multidimensional understanding of a complex situation needs to address its multiple dimensions.

One of the major themes of the prospectus was society's loss of its value base. In the words of Ozbekhan:

The experience of the last twenty to thirty years has shown... that the issues that confront us... may not easily yield to the [analytic and positivistic] methods we have employed in the [past]. The familiar concepts, values, thoughts, and approaches that until very recently served to focus our perceptions, to clarify... what is obscure and unmanageable in our situation are now found wanting... Our problem is that the reality we are beginning to sense [belongs to a] newer order. It is the rationalizing principles of this reality that we are called upon to define (1968).

Hasan recognized that this common value base requires that dialogue has to be built into each design application.

A second major theme of the Club of 'Rome proposal was "the problematique." This term was used to draw a distinction between the well-bounded problems we perceive and articulate and the meta-problem (or meta system of problems) that emerges as a result of the interactivity and inter dependency among these problems. The problematique is roughly equivalent to Churchman's "enormous problems," Ritter and Weber's "wicked problems," and Ackoff's "great gigantic mess." The "49 continuous critical problems" were a first attempt to articulate the problematique.

When Ozbekhan presented the prospectus to the Club of Rome's executive council, he was met with an unsympathetic reception.

Most of those present became increasingly puzzled as to what was its purpose. It was couched in social science jargon which, while it might make sense to his professional colleagues might well cause difficulties to others. Ozbekhan may have had a lot to offer through his academic work but we felt that his style did not match Peccei's vision at this point (Whitehead).

The original conceptualization of the Club of Rome prospectus advocated the position that any attempt at resolving the global problematique founded on traditional elitist, exclusionary, and disciplinary approaches is doomed to failure. The majority of the executive council were locked into a top-down design mindset and rejected the original prospectus as lacking in methodological specificity and rigor. In its place they turned to the systems dynamics approach of Jay Forrester. At this point, Hasan and Aleco resigned from the club. The results of the system dynamics effort were published in *The Limits to Growth* in 1972.

In comparing the prospectus to the Limits to Growth, one can see that Limits is top-down, elitist, and traditional. It selects only five problems that are amenable to traditional and measurement approaches and ignores the 44 others.

In retrospect, one can see that the prospectus presented an architecture within which we could devise a methodology capable of engaging the stakeholders in a dialogical process with sensitivity to their cultural situation and the praxis of their lives. The executive board wanted to do something and the prospectus presented a plan for further dialogue in search of a dialogic methodology.

Aleco has spent his remaining years collaboratively developing such a methodology. A giant leap in that effort came through collaboration with John Warfield. After Aleco gave a talk explaining how he laboriously ascertained the influence that problems have on each other in a particular situation, he was approached by Warfield. Warfield told Aleco that he appreciated what he had done and knew a way to simplify the process. They immediately recognized that they should work together.

The way that Warfield proposed was later developed into the Interpretive Structured Modeling (ISM) methodology. In a span of 25 years, John and Aleco worked together, first at the Battelle Institute, then at the University of Virginia, and George Mason University. They developed, tested, and continuously improved the bottom-up methodology of Interactive Management (IM), which coordinated the Nominal Group Technique with ISM within a carefully modulated learning environment.

The learning environment was designed to enable productive dialogue. It was structured to overcome the unshakeable burdens of that dialogue. A basic burden is the limitation of our working memory that can handle only 7 + or - 2 items at a time. This burden was overcome by selective computer software. The expected group pathologies were controlled by the enforced structure of the dialogue. Unequal power relation within the group were also moderated by the structure of the dialogue. As the structure was refined, its rules were defined. They now exist as the 7 laws of Structured Dialogic Design (SDD).

With Reason and Vision

The Science of Dialogic Design

The intent in this brief book is to make more transparent some of the essential components of the science of dialogic design, with emphasis on its role in the context of designing contemporary human systems, such as national health care systems, human settlements, school buildings, hospitals, and in general organizations of profit and non-profit nature.

The components of the science will be described by employing the evolutionary learning model developed by John Warfield and titled the Domain of Science Model (DOSM) (Warfield, 1986). That model simplified for purpose of this paper is presented below.

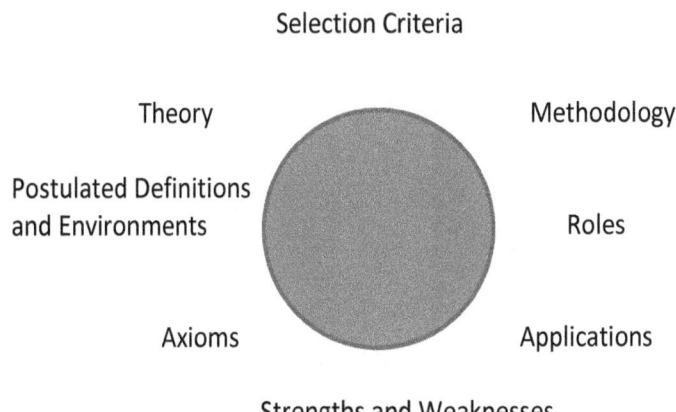

Figure One: Simplified Domain of Science Model

This diagram is to be read clockwise from the lower left starting with Axioms. Reading clockwise we have Axioms --> Definitions --> Theory --> Selection Criteria --> Methodology --> Roles/Environments, Applications --> evaluating Strengths and Weaknesses, which can be extended by possibly reevaluating the Axioms, etc. This sequence provides the frame work for the following presentation.

Foundations

Axioms of Dialogic Design Science

- The Complexity Axiom: Social systems designing is a multi-dimensional challenge. It demands that observational variety be respected when engaging observers/stakeholders in dialogue, while making sure that their cognitive limitations are not violated in our effort to strive for comprehensiveness (John Warfield).
- The Engagement Axiom: Designing social systems, such as health care, education, cities, communities, without the authentic engagement of the stakeholders is unethical, and results in inferior plans that are not implementable (Hasan Ozbekhan).
- The Investment Axiom: Stakeholders engaged in designing their own social systems must make personal investments of trust, committed faith, or sincere hope, in order to be effective in discovering shared understanding and collaborative solutions (Tom Flanagan).
- The Logic Axiom: Appreciation of distinctions and complementarities among inductive, deductive and retroductive logics is essential for a futures-creative understanding of the human being. Retroductive logic makes provision for leaps of imagination as part of value-and emotion-laden inquiries by a variety of stakeholders (Norma Romm and Maria Kakoulaki).
- The Epistemological Axiom: A comprehensive science of the human being should inquire about human life in its totality of thinking, wanting, telling, and feeling, like the indigenous people and the ancient Athenians were capable of doing. It should not be dominated by the traditional Western epistemology that reduced science to only intellectual dimensions (LaDonna Harris and Reynaldo Trevino).
- The Boundary-Spanning Axiom: A science of dialogue empowers stakeholders to act beyond borders in designing symbiotic social systems that enable people from all walks of life to bond across possible cultural, religious, racial, and disciplinary barriers and boundaries, as part of an enrichment of their repertoires for seeing, feeling and acting (Ioanna Tsivacou and Norma Romm).
- The Reconciliation of Power Axiom: Social Systems designing aims to reconcile individual and institutional power relations that are persistent

and embedded in every group of stakeholders and their concerns, by honoring Requisite Variety of distinctions and perspectives as manifested in the Arena (Peter Jones).

These axioms are the fundamental principles of dialogic design science. They have been developed, tested, and revised during forty years of application in actual practice. In the last several years they were codified in an international group effort. We believe that any democratic design should adhere to the spirit of these axioms.

Definitions

It has been observed that participants in dialogue, especially if they come from different backgrounds and have diverse interests, speak different "languages." For example, the word "responsible" for a factory owner might mean getting as much profit as possible for his shareholders; for his employees it might mean offering respect and a fair wage.

An opening task in dialogic design is to have everyone present their ideas in their own words and clarify them in response to queries from other participants. Their wording and definitions are sacrosanct; they may not be challenged. Anyone with a different definition or viewpoint however, is free to present opposing ideas and definitions, which will also be held sacrosanct. After about two hours of this exercise, the participants develop a common language; the tension goes out of the room and the participants are free to go to work.

While definitions in dialogue may vary from person to person, the same cannot be true for key concepts regarding the *process* (not the *content*) of a structured dialogue.

The following eight terminological definitions are inferred from and are complementary with the seven Axioms of the Dialogic Design Science. These definitions establish the foundational language of the science, and are evolving in accordance with the Domain of Science Model (DOSM) of Warfield:

- Dialogue: The engagement of observers/stakeholders in discovering meaning, understanding, wisdom, and actions for designing their social systems by means of structured inquiry in a "colaboratory of democracy."

- Conscious Evolution: The engagement of observers/stakeholders in a colaboratory for the purpose of creating their ideal futures.

- Future: The state of a social system that is significantly different from the state obtained by extrapolating past and present trends.

- Triggering question: A prompt framed by a colaboratory Design Management Team (DMT), in collaboration with the sponsor, for the purpose of enabling observers/stakeholders of the social system to construct high quality observations.

- Elemental Observation: The succinct and content-specific observation by an observer/stakeholder in response to a triggering question during a colaboratory.

- Third Phase Science: All inquiry actions that aim to support observers/stakeholders in constructing high quality observations that make possible the design and implementation of action plans for the conscious evolution of a social system (for an elaboration of the three Phases of science please see the response to a question below).

- Truth: The convergence of the alternative realities (or pluralities) of a group of stakeholders participating in a colaboratory to a consensual, ephemeral, and language-sensitive snapshot of the complex situation they are confronting. This time-and-space-specific snapshot is subject to evolutionary learning by iteration.

- Problem statement: The appreciation by an observer/stakeholder of the dissonance between his/her belief of "what ought to be" and the observation of "what is." These statements of stakeholders with diverse perspectives and life-experiences are value-based and language-sensitive.

Three Phases of Science

Here is a response of Dr. Alexander Christakis (Aleco) to a question about the three phases of science from Dr. Tom Flanagan (Tom):

Tom: I suspect that most people will not see dialogue as a science but rather as an art in common experience. Human beings seem to be hardwired for dialogue. It seems easy to see how dialogue plays a role in the sciences but how did dialogue itself become a science?

Aleco: The contemporary world is like the river of the Greek Philosopher Heraclitus who said: "You cannot step in the same river twice." The way that we step into the river defines the river for us at the time we step in it.

Dialogue became a science when we recognized that different observers have different ways of stepping into that river and that we can select and refine the way that we approach the river as a community of stakeholders. Yes, we do use conversations and discussions in all aspects of daily life, but dialogue is a shared exploration into an unfamiliar river, and this is a specialized aspect of communication. Our methods of inquiry represent a science that lies beneath and supports all of the ways that we collectively construct observations. Dialogue is the infrastructure – a word that means 'beneath the structure' – for all of our collective learning. Dialogue science is a science for learning how to learn together. It is a very deep and inclusive science and it has come of age with the recognition of three major phases in the way that we construct observations and consensual understandings.

First Phase science considered learning as a matter of observing phenomena which are understood to be independent of the observer. In other words, an observer seeing an apple falling from a tree in ancient Athens, Greece in 500 BC, will report the same phenomenon as an observer seeing an apple falling from a tree in New York, today. It was a science deliberately focused on "objectivity" as opposed to "subjectivity." Classical Newtonian physics represents this First Phase science. This science has been dominating the discourse, including policy science, for many generations because of its objectivity and the collection of observer-independent data or facts.

Second Phase science considers learning to be shaped by an interaction between an observer and the entity being observed. Anthropologists understand this, as do business managers. The quality of observation is impacted by the presence of an observer within a community, such as the presence of a boss within the staff break room. In medicine, interactions with clinical staff in clinical environments can impact patient responses, including blood pressure. This is called the "white coat phenomenon." In the physical sciences also, quantum physics recognizes that the observer's perspective impacts the way that fundamental states are understood. For example, one cannot observe the velocity and the position of a particle at the same time. The Newtonian approach to understanding an unfamiliar river works only in First Phase science phenomena, such as apples falling from trees.

Third Phase science recognizes that we all step into the river in different ways and at different times. So it honors everyone's contributions as valid observations. In hindsight it may seem that the leap to recognizing Third Phase science might have been a small step. When a scientist in any discipline makes an observation, that observation is subject to the review of peers within that science – yet the view is not concurrently subjected to the review of scientists in other fields. Why not? The reason is sciences evolve to advance their discipline's understanding of the world, and this view tends to converge upon the beliefs, tools, and prior understandings accumulated into that specific science discipline. Sciences become silos – they become specialized for viewing the world in accordance with the discipline that they view the world.

 What happens when a phenomenon transcends disciplinary silos? How do we look at complex situations like global sustainability or even community infrastructure investment? The understanding of the situation changes with the lens that we use to look at it. For this reason, we have come into an age of Third Phase science. As a global community we are learning how to learn together. This phase of science is not a matter of contesting which view is right and which view is wrong. It is a matter of merging understandings at elemental observations and constructing a new understanding which embraces a larger view of the way that the world should operate as a visionary anticipation.

The future is an unfamiliar river that flows through time. When we step into this river, we must step into it together. If we do not use a Third Phase science form of dialogue, we will not construct a visionary sustainable future that will exist for us all. So Third Phase science is focused on enabling observers to construct superior observations collectively and democratically by employing the science of dialogue.

http://leregardcretois.blogspot.com/2012/02/demosophia-paradigm-as-solution.html.

Theory

Seven Dialogue Laws (Development Years 1995 - 2006)

The application of Dialogic Design Science requires Facilitators of Structured Dialogue to strictly comply with 7 Laws.

Law of Requisite Variety

The Law of Requisite Variety: An appreciation of the diversity of perspectives and stakeholders is essential in managing complex situations. The Law of Requisite Variety is attributed to William Ross Ashby.

Law of Requisite Parsimony

The Law of Requisite Parsimony: Structured dialogue is needed to avoid the cognitive overload of stakeholder/designers. The Law of Requisite Parsimony is attributed to George Miller and John Warfield.

Law of Requisite Saliency

The Law of Requisite Saliency: The relative saliency of observations can only be understood through comparisons within an organized set of observations. The Law of Requisite Saliency is attributed to Kenneth Boulding.

Law of Requisite Meaning

The Law of Requisite Meaning: Meaning and wisdom are produced in a dialogue only when observers search for relationships of similarity, priority, influence, etc., within a set of observations. The Law of Requisite Meaning is attributed to Charles Sanders Peirce.

Law of Requisite Autonomy and Authenticity

The Law of Requisite Autonomy and Authenticity: During the dialogue it is necessary to protect the autonomy and authenticity of each observer in drawing distinctions. The Law of Requisite Autonomy and Authenticity is attributed to Ioanna Tsivacou

Law of Requisite Evolution of Observations

The Law of Requisite Evolution of Observations: Learning occurs in a dialogue as the observers search for influence relationships among members of a set of observations. The Law of Requisite Evolution of Observations is attributed to Kevin Dye. The importance of this law is explained in the following section.

Law of Requisite Action

The Law of Requisite Action: Any action plans to reform complex social systems designed without the authentic and true engagement of those whose futures will be influenced by the change are bound to fail. The Law of Requisite Action is attributed to Yiannis Laouris.

Relationship between the Axioms and the Laws

There has been considerable discussion among the members of the community of practitioners and theoreticians of DDS regarding the relationships between the axioms and the laws of the science. In an effort to clarify these relationships in the context of the Domain of Science Model (DOSM) and the Referential Transparency article: http://dialogicdesignscience.wikispaces.com/file/view/DDSontoDOSM.pdf.

We make more explicit the connections between the axioms and laws in the section below:

The Complexity Axiom: Social systems designing is a multi-dimensional challenge. It demands that observational variety be respected when engaging observers/stakeholders in dialogue, while making sure that their cognitive limitations are not violated in the effort to strive for comprehensiveness (John Warfield).

> *The two Laws deduced from this Axiom are Requisite Variety, and Requisite Parsimony. Significant evidence gathered in the Arena over a period of forty years indicates that these two laws are also* supportive of this axiom. This evidence was reported for the first time in a paper authored by Warfield and Christakis in 1987: John N. Warfield, **and** Christakis, A.N. "Dimensionality," Systems Research 4, pp. 127–137.

The Engagement Axiom: Designing social systems, such as health care, education, cities, communities, without the authentic engagement of the stakeholders is unethical, and results in inferior plans that are unethical, and are not implementable (Hasan Ozbekhan).

The Laws of Requisite Authenticity and Autonomy, and of Requisite Action are deduced from and are supportive of this Axiom.

The Investment Axiom: Stakeholders engaged in designing their own social systems must make personal investments of trust, committed faith, or sincere hope, in order to be effective in discovering shared understanding and collaborative solutions (Tom Flanagan).

The Law of Requisite Authenticity and Autonomy, Requisite Saliency, and Requisite Parsimony are deduced from and are supportive of this Axiom.

The Logic Axiom: Appreciation of distinctions and complementarities among inductive, deductive and retroductive logics is essential for a futures-creative understanding of the human being. Retroductive logic makes provision for leaps of imagination as part of value-and emotion-laden inquiries by a variety of stakeholders (Norma Romm).

The Laws of Requisite Saliency, Requisite Meaning and Wisdom, and Requisite Evolution of Observations are deduced from and are supportive of this Axiom.

The Epistemological Axiom: A comprehensive science of the human being should inquire about human life in its totality of thinking, wanting, telling, and feeling, like indigenous people and the ancient Athenians were capable of doing. It should not be dominated by the traditional Western epistemology that reduced science to only intellectual dimensions (LaDonna Harris and Reynaldo Trevino).

The Laws of Requisite Authenticity and Autonomy, and Requisite Evolution of Observations are deduced from and are supportive of this Axiom.

The Boundary-Spanning Axiom: Stakeholders are empowered to act beyond borders to design symbiotic social systems that enable people from all walks of life to bond across possible cultural, religious, racialized, and disciplinary barriers and boundaries, as part of an enrichment of their repertoires for seeing, feeling and acting (Ioanna Tsivacou and Norma Romm).

> The Laws of *Requisite Authenticity and Autonomy* and *Requisite Action* are supportive of this Axiom.

The Reconciliation of Power Axiom: Social system design aims to reconcile individual and institutional power relations that are persistent and embedded in every group of stakeholders and their concerns, by honoring Requisite Variety of distinctions and perspectives as manifested in the Arena (Peter Jones).

> The Laws of *Requisite Authenticity and Autonomy, Requisite Evolution of Observations,* and *Requisite Variety* are supportive of this Axiom.

It should be emphasized that these laws apply to the process and are guides for facilitators. Participants need to be aware of them only if the dialogue is going off course. With regard to the content of their dialogue, the participants are in complete control.

Spreadthink, Groupthink, and Erroneous Priorities

Professor John N. Warfield

Dr. John N. Warfield, the great pioneer of integrative sciences, uses the term "Spreadthink" to describe the outcome of group dialogue infected with certain behavioral and cognitive constraints. This refers "to the demonstrated fact that when a group of individuals is working on a complex issue in a facilitated group activity, the views of the individual members of the group on the relative importance of problems and/or proposed action options will be literally 'spread all over the map.'"

Moreover, Warfield cautions, "Facilitators who try to bring groups to a majority view or a consensus without the aid of some methodology that resolves the difficulties caused by Spreadthink may well be driving the group to Groupthink, and thus helping to arrive at a decision that lacks individual support and, usually, lacks substance." Groupthink, refers "to the deterioration of mental efficiency, quality of reality testing, and quality of moral judgment that results from in-group pressures. Subject to Groupthink, a group may seem to accept a specific decision; however, if individual group members are confronted with that point of view separately from the group, few members would accept that view as their own."

Aleco Christakis
Most people have heard the phrase "talking the talk, and walking the walk." The standard interpretation of this phrase is that there is a discrepancy between what people say and what people do, *i.e.*, between their words and their actions. Aleco Christakis, one of the principal inventors of the "*Technology of Democracy*" whose unique powers of dialogue facilitation are very much like the "specialist mediators" between the people and their deities that marked the popular Aristocracies of Bronze Age Crete (2000 BC - 1370 BC), has challenged

conventional talk in socio-political systems design, which has become a minefield.

Kevin Dye

The discovery of the "Erroneous Priorities Effect" (EPE) after extensive research by Kevin Dye at the Food and Drug Administration, has led to the recognition that even with good intentions for participative democracy, people cannot collectively walk the talk unless we change the paradigm for languaging and voting. Effective priorities for actions require recognition of influence patterns among interdependent factors. When priorities are chosen on the basis of aggregating individual stakeholder subjective voting that is largely blind to those interdependencies, their efforts are defeated by the EPE.

For an extended discussion on this topic click
http://dialogicdesignscience.wikispaces.com/Erroneous+Priorities+Effect

Action Tree

The influence relations among these seven laws can be plotted using the tools of SDD. In several trials the result of that plotting can be seen in the "action tree" seen below. You can find more explanation of this action tree formation at http://dialogicdesignscience.wikispaces.com/Action+Tree. You might use your own retroductive logic when you are briefed on the SDD methodology to see if you arrive at a different result.

Law of Dialogue 7: Laouris's Law of Requisite Action

This recently discovered law (Laouris & Christakis, 2007) states that the capacity of a community of stakeholders to implement a plan of action effectively depends strongly on the true engagement of the stakeholders in designing it. The accompanying engagement axiom states that designing action plans for complex social systems requires the engagement of the community of stakeholders in dialogue. Disregarding the participation of the stakeholders is unethical and the plans are bound to fail.

Law of Dialogue 4: Peirce's Law of Requisite Meaning

Based on Turrisi, 1997, this law says that meaning and wisdom can only be achieved when the participants search for

Law of Dialogue 3: Boulding's Law of Requisite Saliency

Proposed by Boulding in 1966. It calls for comparisons of the relative importance across ideas proposed by different people. This is secured through the voting process.

Law of Dialogue 1: Ashby's Law of Requisite Variety

Proposed by Ashby in 1958. It calls for appreciation of the diversity of observers (i.e., invite "observers" with diverse views)

Law of Dialogue 6: Dye's Law of the Requisite Evolution of Observations

Proposed by Dye et al., 1999 it tells us that actual learning occurs during the dialogue as the participants search for influence.

Law of Dialogue 5: Tsivacou's Law of Requisite Autonomy in Decision

Proposed by Tsivacou in 1997. This law guarantees that during the dialogue, the autonomy and authenticity of each person contributing ideas is protected, and distinctions between different ideas are drawn as a method of deepening our understanding of each idea.

Law of Dialogue 2: Miller's Law of Requisite Parsimony

Grounded on Miller, 1956 and Warfield, 1988. Emphasizes the fact that humans have cognitive limitations, which need to be considered when dealing with complex multi-dimensional problems. This is secured by the fact that participants are asked to focus on one single idea or one single comparison at a time (structured dialogue)

An Influence Pattern among the Seven Laws of Dialogue

The influence relationships among the seven Laws of the Dialogic Design Science are graphically displayed above in what is referred to the DDS literature as a "Tree of Action." The Tree shows, by means of the arrows propagating from bottom up, the enhancement relationship among the seven Laws. The Law of

Requisite Parsimony, attributed to Miller, is at the root of the Tree of Action. This implies honoring the cognitive limitations of the participants during a dialogue will enhance all the other laws above it. Following the pathways displayed in the Tree, we will reach the culminating top level: Laouris Law of Requisite Action.

Authentic and productive dialogue among the group of stakeholders requires that facilitation teams take care that all seven Laws are enforced. While all the laws are important, an emphasis should be placed on Parsimony in accordance with the enhancement pathway in the Tree of Action. Enforcing "Parsimony" enhances the "Autonomy" of the participants, enabling them to experience the "Evolution" of observations and to learn from each other. As they learn together they begin to appreciate the "Variety" of observations, which leads to a reassessing of their original views about "Saliency," and leads to a deeper understanding of the "Meaning" of the ideas authored by other participants. The understanding of the meanings of the proposals and the whole collaborative design leads to buy-in and the commitment to "Action."

The reader interested to learn more about the science of Dialogic Design and the Tree of Action should consult some of the relevant literature about the science, and in particular (Christakis, 2006; Laouris, 2008; Flanagan & Christakis). If you are interested to construct the Tree of Action by using retroductive logic please visit the Dialogue Game in:

http://globalagoras.org/DialogueGame.pdf

Role Distinctions

There are three key role distinctions discovered during the years 1982 - 2002. They are:

(1) The Context – Design Management Team,
(2) The Content – Stakeholder/Designers, and
(3) The Process – SDD Facilitation Team.

Roles in an SDD Process Application

Understanding your involvement in a problem solving situation is critical for its success. The Structured Dialogic Design (SDD) practitioner is an expert in the theory and practice of the Dialogic Design Science. However, he/she is not to be the expert for problem definition, solutions content, or the designer of the solutions. Knowing how to distinguish between process and content expertise as a facilitator will make you a successful SDD practitioner.

To better help you understand the SDD practitioner roles, let's explore these roles under the following three key Role dimensions: Context, content, and process.

Dimension	Domain	Explanation
The Context: The design situation, what are potential causes for a problem and all surrounding aspects of it. Who are the players (stakeholders, the owner of the problem, whose accountability is it).	Client and SDD practitioner. Here is where a dance, between content and process expertise will take place between the SDD practitioner and the client, the owner of the problem.	The context is what will drive the problem resolution. It is a broad picture of an issue, which will eventually be defined as a "system of problems" or problematique. The SDD practitioner has some knowledge about the type of problem being discussed, and understands about the nature of complex problems. Understands complexity, and will help the client visualize the problem in its contextual situation. Will work with the client in formulating a problem solving question (triggering question). This question is extremely important, because it will guide all the activities prescribed by the SDD process. The triggering

		question will help the client select who are the stakeholders.
The Process: This is where the SDD practitioner exhibits his/her expertise.	This is the exclusive domain of the SDD practitioner.	A successful SDD practitioner would know the process very well, and will know when the process application is appropriate. The SDD practitioner knows what the non-negotiable process aspects are and why they are non-negotiable.
The Content: If the problem has been described as a complex, wicked problem, then it is a STAKEHOLDER dependent situation.	The content for resolving the problem is a stakeholder prerogative.	A SDD practitioner, in his/her facilitator role is never to interfere or interject with participant's perspectives. This is the exclusive domain of the client and the client's stakeholders. The SDD practitioner will only facilitate the content dialogue by employing the science of Dialogic Design.

Methodology

Warfield's Domain of Science Model

At this point, we will review the Science of Dialogic Design by positioning it elements within the larger context of Warfield's Domain of Science model.

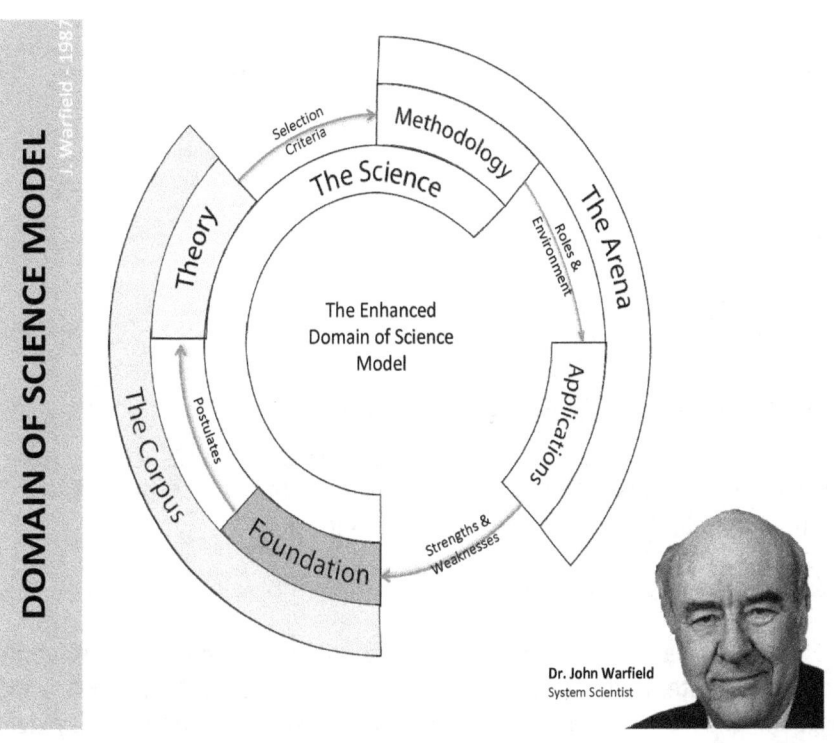

Dr. John Warfield
System Scientist

The axioms belonging in the Foundation domain are used to steer the theory (which includes the Laws of the science). The Theory domain steers the Methodology domain, which steers the Applications. The observations made in the domain of Applications are fed into the Foundation domain, namely the Axioms and the language of the science, and modify them accordingly.

In the case of Euclidean geometry, Euclid was playing in the sand of the Aegean Sea when he discovered his geometry. He conjectured four axioms, such as, you can draw from a point only one parallel line to another line. Euclid was able to deduce from these four axioms all the laws and theorems of his geometry. The Euclidean geometry was very instrumental for the discovery of classical (Newtonian) physics in the 18th century; it became irrelevant, however, for relativistic mechanics in the 20th century. The General Theory of Relativity was founded on the axioms of Riemannian geometry, and not the axioms of Euclidean geometry. All geometries are based on axioms, and the Archimedean geometry of Dialogic Design Science also has its axioms.

Seven Consensus Methods (Development Years 1972 -2011)

(1) Nominal Group Technique,
(2) Interpretive Structural Modeling,
(3) DELPHI,
(4) Options Field,
(5) Options Profile,
(6) Trade-off Analysis, and
(7) Webscope

For further discussion of these methods consult
http://dialogicdesignscience.wikispaces.com/Consensus+Methods+%287%29
The first two consensus method will be discussed in the next sections. The seventh method, the Webscope will be discussed later.

The Nominal Group Technique

This process bears superficial resemblance to brainstorming, but is vastly superior for systemic design purposes. In NGT, responses to a carefully crafted triggering question are written down. Each response is limited to one key idea. Other ideas are expressed in separate responses. In addition, participants present their responses to the group in a round-robin manner. The advantages of NGT used in this way are:

- Participants are given time for thoughtful consideration and participants who are slower or not anxious to be heard are not disadvantaged.
- They retain their autonomy and their control over how their responses are stated.
- They can clarify single ideas without defending them against competing ideas. Any competing ideas may be offered as separate responses.

- They can cluster their clarified ideas with relative ease.
- They can vote for the clarified ideas that they deem most important.

Interpretative Structural Modeling (ISM)

This process takes us beyond individual preferences of importance to collective decisions regarding the influences that individual ideas have on each other. ISM creates an influence map in a problematic situation that allows participants to focus on the root causes of the problem, and to avoid fruitless efforts focused on concerns that may be important, but lack the influence to generate meaningful change. It has been shown that 'important' causes are rarely the 'influential' causes of the problem. This phenomenon has been termed, "the erroneous priorities effect" (Dye and Conaway, 1999).

ISM accomplishes this task by assessing the influences that single ideas have on each other; for example, it asks "If we make progress on addressing cause A, will that significantly help us to address cause B?" This potentially overwhelming task is made easier through the use of special software that keeps track of the individual decisions and plots the transitive logic among them. The software creates a map of the influences and even portrays them in the form of an 'influence tree'.

In the whole process of Structured Dialogic Design participants draw on resources beyond mere cognitive wisdom. They assess their feelings, unconscious hunches, intuitions, and sense of solidarity to create *prescriptions* for their future behavior. In doing this, they transcend the *descriptive* boundaries of first and second phase science.

For a further elaboration of six of the seven Consensus Methods see *How People Harness their Collective Wisdom and Power:*
http://www.amazon.com/gp/product/1593114818/ref=ed_oe_p/104-7420823-4256710?%5Fencoding=UTF8

For a more thorough discussion of the Consensus Methods the interested reader should consult:
Warfield, John N., and A. Roxana Cardenas. A Handbook of Interactive Management. Ames, Iowa: Iowa State University Press, 2d, 1994

Seven Language Patterns (Years 1970-1989)

SDD utilizes seven language patterns in analyzing a situation, understanding it, and prescribing corrective action:

1. Elemental observations are the simple reports and opinions of all stakeholders in a situation. The reports and opinions of ordinary people are of value because they are the experts when it comes to their own reality.
2. A *problematique* is a very complex troublesome situation. It is also the expressed result of an SDD session that presents a resolution to that situation.
3. An influence tree pattern is the graphic representation of a *problematique.*
4. An options field pattern is the graphic representation of proposed solutions of the systems sorted according to their dimensionality.
5. An options profile/scenario pattern is a selection of options from the options field fits with a specific scenario or pattern.
6. The superposition pattern combines the Influence Tree among issues with the action options of the options profile in a unified graphic representations.
7. And the Action plan pattern embodies the assigned responsibilities, time lines, interaction patterns, and monitoring activities decided on by the participants.

For further discussion of language patterns, see
http://dialogicdesignscience.wikispaces.com/Language+Patterns+%287%29

Graphic Language Patterns

The graphic language patterns of the Archanesian geometry, shown below, are used by the stakeholders in Co-Laboratories of Democracy to construct translatable graphics relevant to their specific social system designing situation. The graphic below portrays the steps in each stage of inquiry.

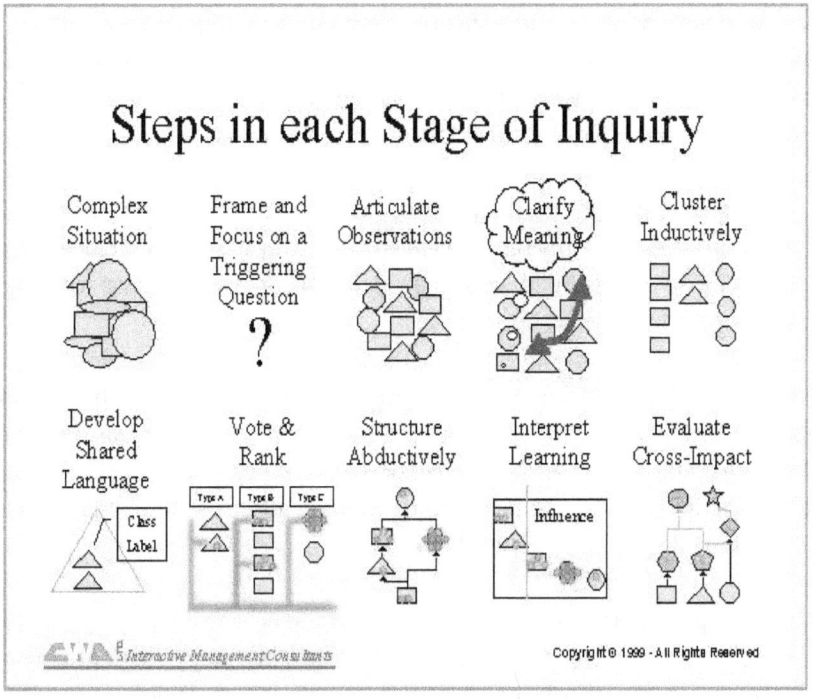

Image of the Steps in each Stage of Inquiry (The Archanesian Geometry)

Step (a) is not really a step. It is the complex situation that Structured Dialogue is asked to address. It consists of many interrelated institutions, ideas, cultures, economic constraints, etc. This hodgepodge is investigated with the goal of framing triggering questions.

In step (b), the triggering questions frame the context of the dialogue. A sample triggering question might be: "What are the strengths of this organization and what is hindering its progress?"

In response to this question, the participants articulate their ideas in their own words to the full attention of the other participants, step (c). Their words are posted on a wall and everyone agrees not to alter them. In a second round robin, step (d), participants respond to questions asking them to clarify (not to alter) their ideas, and are given the opportunity to respond to questions in order to explain their meaning.

This methodology authenticates each person irrespective of his or her education level or position of power. It produces a palpable reduction of tension. People seem surprised as they are being heard, perhaps for the first time, in important policy-making matters.

In step (e), the participants collaborate to inductively cluster the observations they have made. Then in step (f), they agree upon labels for the clusters. These steps build a a shared language and a sense of shared competence within the group.

In step (g), participants rank these clusters according to their relative importance. This step brings into sharp relief the different priorities and values within the group. In the ensuing discussion, parties come to understand where their co-participants are coming from, which leads to a respectful working relationship based on defined mutual interest.

In step (h), participants explore relationships among the observations and construct a tree of relational influences. In this step, they order their observations in paired comparisons asking whether A really influences B, and vice-versa.

Finally in steps (i) and (j), the stakeholder/designers examine the "tree of meaning" they have constructed, with computer assistance. As a group, they analyze and interpret the cross-impacts existing among the observations they have made.

In these ways, step-by-step, Structured Dialogue progressively clarifies the situation and opens the way to greatly enhanced decision-making and action-planning. In addition it:

- Authenticates every stakeholder/participant;
- Elicits ideas and points of view from all stakeholders;
- Moves toward effective consensus;
- Elicits and deals with the different priorities of stakeholder participants;
- Equalizes power relations among the stakeholders;
- Goes beyond identifying factors that are important, to specifying those that are most influential in achieving goals.

Colaboratories achieve this enjoyable and rewarding experience through a variety of methods:

- They structure the dialogue to achieve maximum openness and efficiency within time restraints;
- They employ computers and software (CogniScope 3) to pace and track the discussion, thus relieving participants of the need to take notes and freeing them to focus on the content of the discussions.
- They display participant contributions and post them on the walls of the colaboratory in real time,
- They distribute progress reports in real time.
- The Interpretive Structural Modeling (ISM) software calculates and displays the influence relationships among ideals/obstacles/actions in real time and generates a graphic "Tree of Meaning" displaying those influences.

Colaboratories of Democracy

A Science of Dialogic Design needs to meet two competing criteria. It must present a clear and convincing theory backed by sound and detailed argument. Meeting this criterion requires a large amount of clarification that can generate confusion, boredom, and resultant rejection of the science in practice. The previous text has dealt mainly with the bothersome details, which are the concern of theoreticians and practitioners of SDD.

On the other hand, the science needs to be practical. It must be readily understood and easily implementable. People should find it easy and rewarding to practice. This practical aspect of SDD, which we have named Colaboratories of Democracy, is the focus of this section.

Jeff Dietrich portrays this journey very simply:

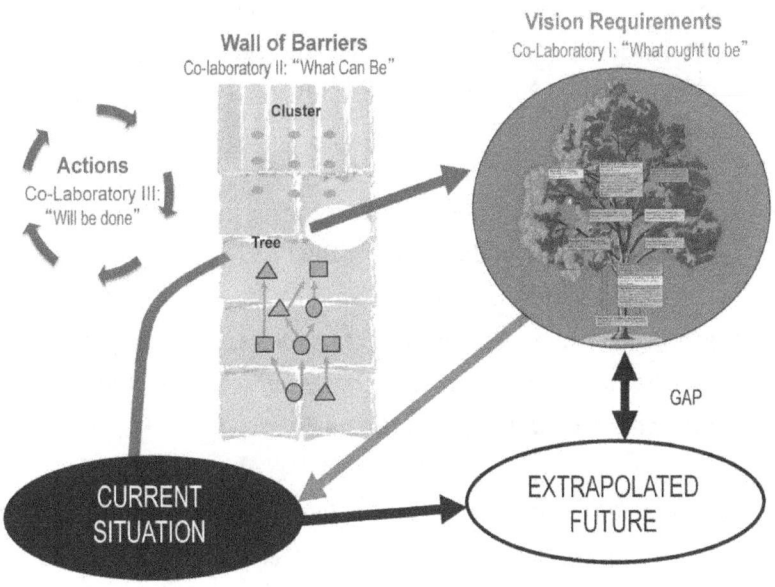

SDD for Futures-Creative Action Scenario Construction

The graphic identifies the three stages of Structured Dialogic Design: Idealizing the desired future; identifying and prioritizing the obstacles that stand in the way of achieving that future; and choosing the actions that will allow us to overcome those obstacles. Note that the alternative procedure of directly extrapolating the future is rejected because it always results with simplistic changes on the mere fringes of problems.

Archetypes of Colaboratories of Democracy:

An archetype (pronounced /ɑrkɪtaɪp/) is an original model of a person, ideal example, or a prototype upon which others are copied, patterned, or emulated. From over thirty years of applying Dialogic Design Science in the Arena we are able to distinguish six Colaboratory of Democracy Archetypes. These are:

1) Type A: Diagnosis of the Problematique - complex primarily through vaguely defined and intensely interacting mega-trends. This type is the most frequently applied in the Arena. It is used for diagnosing a complex problem situation and for discovering the deep drivers for the purpose of initiating a strategy for resolution.

   An example of such a problematique is the dysfunction of our educational system as portrayed in the following URL: http://www.youtube.com/watch?v=zDZFcDGpL4U. An appropriate triggering question should we want to address this entangled mess might be: *"What are issues to be collectively addressed in transitioning to a contemporary paradigm for education?"*

   An example of a less immense problematique was actually carried out with a group of high school students at risk of dropping out of school in Michigan. A report and video on this colaboratory can be seen at: http://cwaltd.wetpaint.com/page/Root+Cause+Mapping+with+Everett+Students. This Archetype has been historically very popular. It was used frequently by a team consisting of John Warfield, Roy Smith, Scott Staley, and Ben Broome with a group of Executives and Engineers of the Ford Motor Company in the 1990s.

The largest number of participants with this Archetype was implemented by Robert McDonald in 1983. He engaged a group of 250 private forest landowners/stakeholders in the USA, under the sponsorship of the Under Secretary of the US Department of Agriculture.

A variation of this archetype that has been applied successfully in some cases, is to allocate about two hours, after the group has constructed the *Problematique,* in engaging the participants in small group work focusing on actions for addressing alternative pathways of the *Problematique.* Following the completion of the small team work, the small teams make brief presentations at a plenary session. The pathway-focused action scenarios proposed by the small teams are useful to the organizational entity for prioritizing the actions for addressing the drivers of the *Problematique.*

2) Type B: Reconnaissance - complex primarily through unexplored situations and unexamined intentions.

This type is applied when we need to gather information and intelligence from a variety of stakeholder perspectives about a complex situation, which is challenging but not necessarily a burning issue. An example of such an application is the recent inquiry (November, 2010) for improving the theory and practice of the Science of Implementation, by engaging a group of theoreticians and practitioners of the science, together with a group of customers in the state of Michigan. For more details visit:

http://mi3implementationscience.wikispaces.com/.

Another example of this archetype is an application in Tokyo, Japan in 2005, with Laura Harris of Americans for Indian Opportunity (AIO) and Kate Cherrington of Advancement of Maori Opportunity (AMO), being the facilitators of a colaboratory with participants from a variety of Asian tribes and cultures: http://quergeist.net/Christakis/interview-Christakis-LaDonna-Harris-p4-5-Summer-2005.pdf.

3) Type C: Long Range Action Scenario Construction - complex primarily through uncertain futures.

This type is employed to explore alternative futures derived on the basis of the extrapolation of past and present trends and events. It enables a group of stakeholders to converge to a consensus action scenario for implementing changes to the extrapolated future. This archetype was designed and implemented originally by Kevin Dye.

A good example is provided by the report below exploring alternative energy efficiency futures for the Pacific Northwest Region of the USA: http://sunsite.utk.edu/FINS/loversofdemocracy/SDDP_Reports/NEEA FinalReport.pdf

Another example of this Archetype was implemented in Mexico in 1994, with the engagement of a panel of about 20 International and Mexican experts (including Hasan Ozbekhan, Erwin Laszlo and John Warfield) on forecasting trends and event to the year 2020. The panel explored international and national alternative futures on a stage in front of an audience in an Amphitheater of about 1,000 Mexican students and citizens. Reynaldo Trevino was the leader of this event, which was conducted simultaneously in Spanish and English. Aleco Christakis was the Facilitator in English, and Carlos Flores in Spanish.

4) Type D: Futures – Creative - complex primarily through unvoiced transformational hopes. This type is applied when we want to transcend the past and present trends and to create an ideal future for a social system.

There are many examples of this type of applications over the last thirty years of practicing the science in the arena. Two recent applications in Michigan, one focusing on idealizing the learning of math by all students, which includes a virtual engagement of stakeholders, is reported in: www.mimi2010.wikispaces.com. A second application focuses on Universal Design for Learning for All students by engaging approximately 30 stakeholders in three f2f Colaboratories of Democracy:
http://attachments.wetpaintserv.us/HglOgmol%2BoYSt4dVzfPVyw%3 D%3D923350.

Another good example of the application of this Archetype with a group of stakeholders of Region 3 of Michigan is discussed in a video presentation by Aleco Christakis to the regional stakeholders. The presentation is made while they were meeting to agree on the cascade of activities from the regional level to the building level for implementing the Action Plan they constructed collaboratively after completing two co-laboratories focusing on Region 3:
http://remc.adobeconnect.com/p56406963/?launcher=false&fcsConte nt=true&pbMode=normal.

In July of 2010 this Archetype was applied in Cyprus for the purpose of establishing a platform for symbiosis between Israelis and Palestinians: www.youtube.com/watch?v=E7NSjhZno80. This particular Archetype has been applied extensively in Cyprus and other countries of the European Union by Dr. Yiannis Laouris and his team http://fwcis.blogspot.com.

5) Type E: Collaborative Action Agenda - complex primarily through the number and diversity of essential collaborators. This two-day CoLaboratory Archetype, with the participation of up to 30 stakeholders, is applied when we want to engage a group in a short-term collaborative action agenda for addressing a pressing issue, which might entail a significant reallocation of resources and a change in policy direction.

An example of this Archetype was a convening in 2003 a group of Medical Nefrologists focusing on addressing the issue of Chronic Kidney Disease (CKD), for which the USA Federal Government spends nearly $20 Billion per year:
http://dialogicdesignscience.wikispaces.com/file/view/CKDFinalReport.pdf

This particular Archetype has been applied extensively to address health care and patient safety related policy issues in the USA during the decade of 2000-2010, primarily under the sponsorship of the National Patient Safety Foundation. A scientific paper describing this application was published in the journal of Nefrology with the principal author Being Dr. Tom Parker who was the Broker for this application.

6) Type F: Root Cause Analysis _ complex primarily through the merging of observer-independent and observer-dependent data.

In March 2004, CWA Ltd., in collaboration with The Great Lakes Area Regional Resource Center (GLARRC), and the Michigan Department of Education, Office of Special Education and Early Intervention Services (OSE/EIS), designed and conducted a root cause analysis co-laboratory with the engagement of thirty stakeholders. The participants to the colab were representatives from the community of practitioners in the field of a monitoring process called Continuous Improvement Focused Monitoring (CIFM). These practitioners were responsible, among other things, for implementing for the state of Michigan the No Child Left Behind (NCLB) legislation, passed by the US Congress in 2002. The participants were initially engaged in a series of colabs for the purpose of designing the CIFM process relevant to their situation, which they will then have to implement in the field with school districts throughout the state.

After the designers completed the design of the CIFM process, it was decided to conduct a "root cause analysis colab" with the engagement of the same group of designers/participants. The purpose of this particular colab was to try to anticipate any factors that might inhibit the successful implementation of the CIFM process in the field. The intention was to conduct an anticipatory root cause analysis, as opposed to one that is the result of an existing systemic problem(s). For more details on this case please visit: http://cwaltd.wetpaint.com/page/Root+Cause+Analysis.

7) Type G: Evaluation through Indicator Rating:

Recently, some innovative alternative application models have emerged. These models are currently being tested in the Arena for gathering evidence. One such model is being developed by Jeff Diedrich. It involves using a panel of experts to determine weights to be assigned to fifty-three Assistive Technology (AT) Indicators, which have been developed by this panel and classified in eight distinct categories. Those weights will be used, together with other metrics at the local level, to assess the performance of an educational agency in the context of delivering AT services to its community of stakeholders. The other innovative application is being developed by Yiannis Laouris in the context of the International Conference this May of the Hellenic Society for Systemic Studies (www.HSSS.gr). It involves the engagement of a variety of stakeholders in making risk assessments for public policy initiatives of the European Union.

For a matrix showing more details about the Colaboratory Archetypes please visit: http://dialogicdesignscience.wikispaces.com/Matrix+of+Co-Laboratory+Archetypes.

Summary of Distinctions among Archetypes of Colaboratories of Democracy:

Type A: Diagnosis of the *Problematique* - complex primarily through vaguely defined and intensely interacting mega-trends

Type B: Reconnaissance - complex primarily through unexplored situations and unexamined intentions

Type C: Long Range Action Scenario Construction - complex primarily through uncertain futures

Type D: Futures – Creative - complex primarily through unvoiced transformational hopes

Type E: Collaborative Action Agenda - complex primarily through the number and diversity of essential collaborators

Type F: Root Cause Analysis - complex through the merging of observer-independent and observer-dependent data

Type G: Evaluation by Indicator Rating - complex through the diversity of indicators measuring a social or natural phenomenon.

If you are interested in downloading a brochure describing colaboratories of democracy in order to share with others please visit:
http://www.harnessingcollectivewisdom.com/pdf/How_Co-Laboratories.pdf

To explore the relationship between the three types of colaboratories and the seven Archetypes it is interesting to study the diagram below:

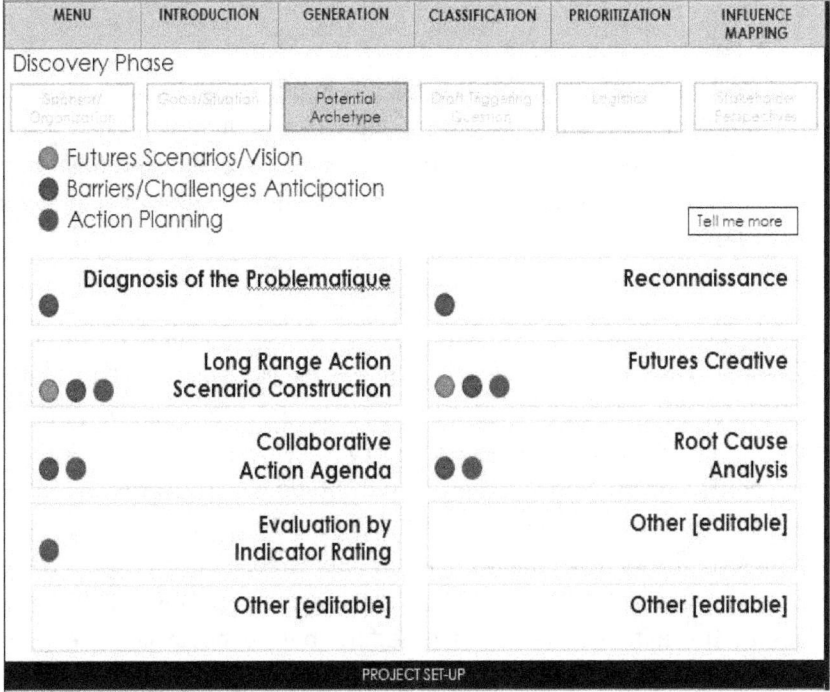

Anticipating the Challenges to the Vision of a Bottom-Up Democracy

The following was an online colaboratory in *problematique* mode that was conducted in asynchronous and synchronous formats. It engaged participants from Australia, Asia, Europe, Africa, and North America. It demonstrates how colaboratories can bring sense and indicate paths to effective action among vaguely defined and intensely interacting mega-trends.

President Obama's Vision

Soon after Obama's victory in the 2008 election, activists noted that Obama's platform was strong in its advocacy of Open Government; it indicated his "interest in employing a bottom up style of decision making" (Huffington Post 11/30/2008).

Then,

On January 21, 2009, his first full day in office, the President issued a Memorandum on Transparency and Open Government, in which he called for recommendations that make the Federal government more transparent, participatory, and collaborative (NAPA memorandum 6/1/09).

The beginning stages of this effort had, however, begun soon after the November election. The National Coalition for Dialogue and Deliberation (NCDD), for example, developed seven core principles for effectively building mutual understanding, affecting policy development, and inspiring collaborative action among citizens and institutions. (To learn more about these core principles, see www.thataway.org/2009/pep_project.)

Members of the Agoras' team were vitally interested in these efforts. We felt we had something vital to contribute to the discussion. We submitted our ideas to the national brainstorming effort, but to no avail. We rather anticipated the rejection of our efforts due to the nature of the national research effort. We will briefly discuss the national effort before we present our colaboratory response to the presidential vision of a bottom up democracy.

The National Exercise

The vision of the national effort was,

To invert the policymaking process by enabling informed public dialogue to inform policymaking at the front end. The collaborative three-phase process opened up tremendous possibilities for real-time innovation. People were invited to:
1. Brainstorm—share ideas on how to make government more open, participatory and collaborative, discuss and vote on the ideas of others;
2. Discuss—dig deeper on the ideas and challenges identified during the Brainstorm phase; and

3. Draft—collaboratively craft constructive recommendations for an Open Government Directive (NAPA memorandum).

The loosely formed question for the weeklong Brainstorming guaranteed an immense, disorganized mass of recommendations of which "legalize marijuana" was a high vote getter. The further progress of this massive exercise was relegated to an expert advisory panel which considered three overarching questions:

- What were the general observations of the week-long discussion?
- What were the most important themes to emerge across the ideas? Where did ideas "cluster" or nodes form?
- Which ideas submitted to the brainstorming present potentially actionable next steps that should be considered for further discussion in Phase II and were viewed favorably by the participant community? Note: the vote totals received were a contributing factor but not the only means for determining the potential value of any specific idea (NAPA memorandum).

The expert analysis of this data consumes 15 pages and may provide some minimal guidance to the Administration's efforts toward participatory government. On the whole, however, this exercise has to be discouraging because of the unstructured and ineffective methodology employed.

The SDD Approach

The Agoras' team took a more practical approach to the presidential call for recommendations. An international team of practitioners of the science of Structured Dialogic Design (SDD), from eight different countries, met on line and off and during December and January. The 15 participants were chosen based upon their

1. Familiarity with the use of the Web
2. Knowledge and experience with the theory and practice of SDD,
3. Interest in the practice of participative democracy on a global scale.

The Webscope wiki was created during the evolution of the Internet. The internet was changing into Web 2.0 so that users no longer just read what was online, they were able to easily write online, which meant that the average person could create a web site that anyone in the world could access. The web became more social with blogs, wikis and sharing photos and videos. In 2007, Aleco approached Gayle Underwood with his interest in getting the SDD process online so participants could interact in their own time and own space. Aleco and Gayle chose a wiki platform and designed the pages of the wiki to fit the SDD process as closely as possible. A 'Generic' wiki was created (http://genericwiki.wikifoundry.com/) to be used as a template for generating more webscope wikis that could be customized for each problematique.

Six Essential Steps for Engaging Stakeholders

SIX ESSENTIAL STEPS FOR ENGAGING STAKEHOLDERS IN VIRTUAL STRUCTURED DIALOGUE CO-LABORATORIES

Step 1: Select a group of stakeholders interested in an issue.

Step 2: Identify a member of the group that is willing to play the role of the Leader in terms of availability to interact directly with the Knowledge Management Team (KMT), which is responsible for managing the structured dialogue via the virtual Colab.

Step 3: The team leader frames, in collaboration with the KMT, a "triggering question" appropriate for the issue selected for focused and open dialogue among the team of stakeholders.

Step 4: All stakeholders participate in a 1-hour Webinar that will orient everyone to the use of the Wiki.

Step 5: Team members are engaged "asynchronously" for approximately three hours in two rounds of "generative dialogue," and two rounds of valuation of the proposals. The three hours of group dialogue can be spread over a week or more, depending on the issue and the availability of the participants to interact asynchronously.

The group responds to the triggering question by entering their statements into a customized website. Team members then vote on the relative importance of the statements they have generated.

Step 6: Team members then engage synchronously for approximately three hours in a strategic dialogue responding to pairwise questions transmitted vocally and displayed onscreen via the Internet. The result is a tree of influence that identifies the leverage points for successful implementation

Members of the Agoras Team and Time Zones as posted in the Wiki

This team worked together to discover the roadblocks facing President Barack Obama in realizing his vision of a bottom-up democracy. The anticipatory stages focused on factors that were likely to inhibit the actualization of Obama's vision. Given that context, great care was given to formulating a proper triggering question.

The Triggering Question

"In the context of Obama's vision for engaging stakeholders from all walks of life in a bottom-up democracy employing Internet technology, what factors do we anticipate, on the basis of our experiences with SDD, will emerge as inhibitors to the actualization of his vision?"

The team participated asynchronously (different places at different times employing the Webscope wiki http://obamavision.wikispaces.com) for the beginning stages of SDD (December 1-12). The group then met synchronously (different places at the same time using web conferencing) for the influence structuring sessions December 13 and January 31).

Responses

The participants generated 59 inhibitors (round one) and clarified them with 60+ messages (round two). They gathered the inhibitors into 13 clusters (round three) and then each participant cast 5 votes for the inhibitors they deemed most important (round four). Among the 59 inhibitors chosen in round four, 15 received 2 or more votes.

Some examples of the inhibitors and their clarifications in the words of their proposers were:

Inhibitor 6: Risk of losing stamina to sustain 'resource hungry' participative processes (rsmith135):

I have experienced a number of industry and civic participation programmes fizzling in the long term, when leaders lose the stamina to sustain the relentless demand on resources required by participative processes. (I am in favor of participative processes. Sadly leaders are rarely willing to spend resources on getting it right the first time, and are willing to squander resources by starting over again after they got it wrong.)

Inhibitor 7: Confusions leading to exclusion of stakeholders with different lingual and cultural background (Heiner). *The dream that all people speak the same language is great – but unreal – as what is said and written and what is meant depends on the cultural setting and situational context. Speaking not the "right" language and not the common meaning causes exclusion. And even worse, when people think they know and understand – but do not – the trouble is there. So alienation starts with not asking what stakeholders mean in certain contexts. Doing it not F2F (face-to-face) reduces the chances to get the differences that matter (gesture, smiles, frowns, or any kind of body language or silence as a way to communicate and react).*

Inhibitor 8: Technical-technological exclusion (Heiner): *The discussion of inclusion and exclusion in our modern-times is extensive. Internet makes exclusion less visible, so stakeholders start guessing if they are left out (or not). They might realize that they miss a certain feature or functionality, but typically they are just left out of the game – feel alienated and so segregation starts.*

Inhibitor 9: Overwhelming variety of individual concerns (phjones): *A significant factor inhibiting the realization off democratic participation is the overwhelming variety and volume of the concerns individuals will raise in an open-ended context. Without a clearly framed scope for engagement, random participants (in an Internet environment) will assert claims based on their personal and closely-held concerns. In a bottom-up approach without a clear frame of reference, these claims may not map to a common ground of understood issues. The result is a confusing problematique and inability to resolve differences between XXXX and YYYY.*

The 15 inhibitors that received 3 or more votes were:
2: (4 Votes) Risk of excluding disadvantaged people (rsmith135)
8: (3 Votes) Technical-technological exclusion (Heiner)
14: (3 Votes) Insufficient attention given to facilitator capacitation (Norma Romm)

18: (3 Votes) Scalability of discussion technology (paulhayes)
23: (3 Votes) Social contract overload (tom_flanagan)

In round 5, the participants, using the Webscope wiki, engaged in synchronous interaction to construct a relational map displaying the influences among factors of higher relative importance, as determined from the voting results of Round 4. This "root cause map" draws distinctions between factors that exert strong leverage, appearing at the roots of the relational tree (map), and factors that are less influential appearing at the branches of the tree. Drawing these distinctions is critical in solving the roadblocks to actualizing bottom-up democracy. Without diagnosing the roots of the issue stakeholders run the risk of allocating resources to the symptoms and not the causes of the problem.

Influence pattern among Inhibitors

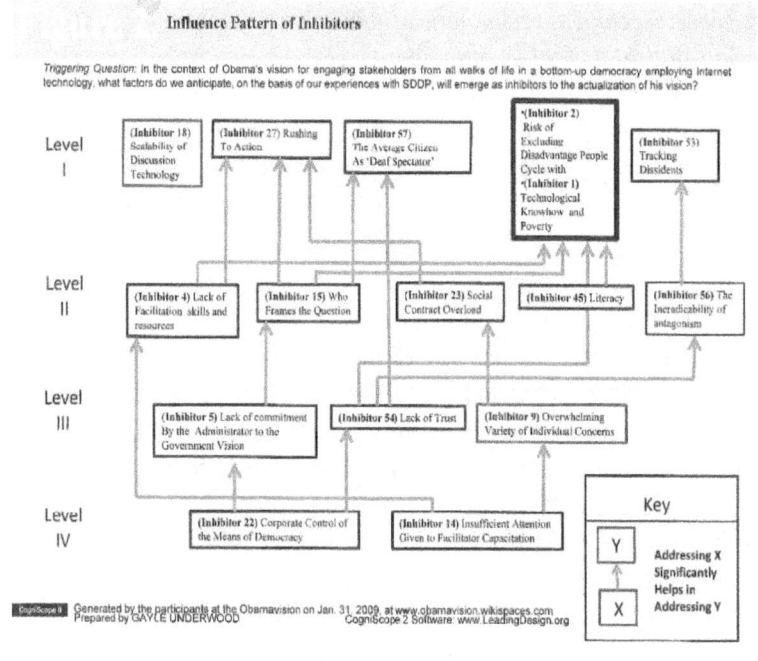

The deepest drivers, the inhibitors with the greatest leverage, identified by the participants were: (**22.** Corporate control of the means of democracy) and (**14.** Insufficient attention given to facilitator capacitation).

It should be noted that the most popular Inhibitor (**1.** excluding disadvantaged people) is not very influential; efforts would be victims of misplaced priorities. Notice also that the overall goal of enabling pathways for many, many voices with many, many ideas to flow in an orderly fashion toward the highest summits of national thinking, while extremely important, is not a most influential driver in its achievement.

Corporate control of the means of democracy emerged as the most influential inhibitor to the realization of the bottom-up democracy vision. In other words, in accordance with the majority vote of the team, the initiative that exerts the highest leverage in inhibiting the practice of bottom-up democracy is the corporate control of the means of democracy.

Insufficient attention given to facilitator capacitation emerged as the second most influential inhibitor. Dr. Tom Flanagan provides a narrative regarding this inhibitor.

> "It is perhaps no great surprise that when a panel of systems scientists from across the globe pull their heads together around challenges that President Obama is likely to face...the most influential factor underlying the success of such an outcome was judged to be the commitment that government leaders and agencies actually hold in supporting a grassroots effort. The global design team phrased this as 'insufficient attention given to facilitator capacitation.'

Overall Evaluation

In his appraisal of the Obamavision findings, Tom Flanagan offered the following observations:

- We are rarely sure if any of our sentiments move forward with our elected representatives.
- The way that they make tradeoffs to capture power for greater leverage (on our behalf) is an entirely non-transparent process.

- This lack of transparency provides a basic rationale for President Elect Obama to consider opportunities for doing something differently.
- The "something differently" in this instance, is a parallel path that comingles ideas with existing legislative processes.
- To redesign a system for managing information, there are few instances where one system would suddenly replace another. With two competing processes in place, there is ample opportunity for an incumbent process to passively starve or actively sabotage an emerging new practice.
- Building the political will among high ranking political officers and institutional administrations is a daunting challenge.
- If the political will is found to assure "capacitation" of a process that facilitates a grassroots contribution to national policy, then this "capacitation" can lead to an evolution within which the overwhelming variety of individual concerns across the nation are brought to the planning table; and within which democratic skills and resources become progressively more available to serve the peoples of the nation.
- Corporate interests can show themselves through the suprahuman power of the corporate voice.
- Corporations can control the means of practicing democracy through their messages and also through their economic control over the media.
- Fundamental changes in the nature of the power of corporations may be needed this stranglehold, which nourishes the interests of corporations and starves the interests of people.
- Reduced corporate control will also open the ears of the "deaf spectator" citizen who otherwise would see the contest for outcomes too heavily tilted to the corporate sector. This could herald a new era of citizen will toward participatory democracy.

Full participation is an essential goal, even if this goal can be achieved only in degrees.

In conclusion, the system of factors related to grassroots participation reflect back on demonstrated acceptance, use of, and support for democratic process at the highest levels of government.

Evaluation of the SDD Methodology in this Effort

This colaboratory mostly praised SDD and the Webscope wiki; it had, however, some liabilities. In 2009, the Webscope wiki was put together "with baling wire and duct tape." In addition, back then the software had no ability to scale up to handle hundreds and even thousands of participants. Back then, we could not offer a convincing alternative to the standard approach employed by the national exercise.

It is particularly interesting that while the global team did nominate the idea of "scalability of discussion technology" as being a central concern, the global team did not map this factor in the system of factors that constitute the essential challenge that the president elect faces in implementing a bottom up style of decision making. This is perhaps due to the fact that there are "missing links" of technology that impede Internet scalability.

One strategy is to supply missing links by conducting many small group sessions. There is at this time a practice called Study Circles (www.studycircles.org) and there is a practice within the participatory dialogue arena that is called Deliberative Dialogue (www.kettering.org). In both of these cases, scale is achieved by multiple reenactments of smaller group practices. The scalar phenomenon is related to practicing the same approach, albeit not at the same time and in the same place. The experiential impact is that participants develop some trust for certain types of processes. If the process does enable authentic democracy and is strongly protected wherever it is used, it might inspire citizen confidence. Reasonable groups who pull together diverse perspectives such as they themselves have done, are likely to draw similar if not identical conclusions.

Accepting this as a hypothesis involves a huge leap: we do not have many highly codified processes for assuring complex problem solving through participatory democracy. SDD has the potential for supplying an effective and tested methodology that can work in this distributed model.

Even where we do have powerful socio-technologies for assuring participatory democracy at a grassroots level, we do not have any valid reason to believe that such processes are used in the highest halls of government. Efforts to reinvent democracy will have to include campaigns to employ SDD in multiple levels of government planning.

A major missing link is an effective methodology, which can scale up to involve hundreds or thousands of empowered participants at one time. Now SDD is about to face that challenge. In May 2015, Yiannis Laouris and his team at the Future Worlds Center presented their planned protocols for enabling an Internet capable Cogniscope3 for use with hundreds of participants. In 2015 with the development of CogniScope3, major steps toward scalability are almost here.

Resources for Further Information and Examples of Virtual Colaboratories

For further information on the Obamavision SDD, see www.obamavision.wikispaces.com.

You can watch a new documentary Dialogue Beyond Borders here: https://www.youtube.com/watch?v=WoATEmPVRoc&feature=youtu.be. It features Arab-Israeli and Turkish-Greek Cypriot observations during SDD peace negotiations.

For an example of hybrid mixed presence, i.e., both f2f and virtual via the Internet, see a co-laboratory focusing on Michigan's Integrated Mathematics Initiative: http://mimi2010.wikispaces.com.

Other cases focusing on Assistive Technology are displayed in:

http://mits-at.wikispaces.com/

www.region1atproject.wikispaces.com

http://region3atproject.wikispaces.com/

References

Christakis, A.N (1988) Th3e Club of Rome Revisited in *General Systems. W.J. Reckmeyer (ed.),* International Society for the Systems Sciences, vol. xxxi, pp 325-38.

Christakis A.N. and Bausch, K.C. (2006). *How People Harness their Collective Wisdom and Power to Construct the Future in Co-Laboratories of Democracy.* Information Age Publishing, Charlotte, NC.

Flanagan, T.R. and Bausch K.C. (2011). *A Democratic Approach to Sustainable Futures.* Ongoing Emergence Press, Riverdale GA.

Flanagan, T.R. and Christakis, A.N. (2010). The *Talking Point: Creating an Environment for Exploring Complex Meaning.* Information Age Publishing, Charlotte, NC.

Meadows, D.H., Meadows, D., Randers, J. (1972). *The Limits to Growth.* New York: Universe Books.

Ozbekhan, H. (1969). Towards a general theory of planning. In E. Jantsch (ed.), *Perspectives of planning.* Paris: OECD Publications

Peccei, A. (1969). *The Chasm Ahead,* Toronto: Macmillan.

Warfield, J.N. "The Domain of Science Model: Evolution and Design," *Pr4oceedings of the Society for General Systems Research,* Salinas, CA: Intersystems, 1986, H22-H59.

Warfield, J.N. and Cardenas, A.R. (1994). *A Handbook of Interactive Management,* Ames IA: Iowa State University Press

Whitehead, J.R. (date unknown) *A Brief History of the Club of Rome: A Summary and Personal Reminiscences.*
http:// www.sympatico.co/drrennie/DACORhis.html.